YOUR KNOWLEDGE HAS VALUE

- We will publish your bachelor's and
  master's thesis, essays and papers

- Your own eBook and book -
  sold worldwide in all relevant shops

- Earn money with each sale

Upload your text at www.GRIN.com
and publish for free

Bibliographic information published by the German National Library:

The German National Library lists this publication in the National Bibliography; detailed bibliographic data are available on the Internet at http://dnb.dnb.de .

This book is copyright material and must not be copied, reproduced, transferred, distributed, leased, licensed or publicly performed or used in any way except as specifically permitted in writing by the publishers, as allowed under the terms and conditions under which it was purchased or as strictly permitted by applicable copyright law. Any unauthorized distribution or use of this text may be a direct infringement of the author s and publisher s rights and those responsible may be liable in law accordingly.

Imprint:

Copyright © 2018 GRIN Verlag
Print and binding: Books on Demand GmbH, Norderstedt Germany
ISBN: 9783668885776

This book at GRIN:

https://www.grin.com/document/453922

Deapon Biswas

How to Make a Triangulum and Special Combination Series

GRIN Verlag

GRIN - Your knowledge has value

Since its foundation in 1998, GRIN has specialized in publishing academic texts by students, college teachers and other academics as e-book and printed book. The website www.grin.com is an ideal platform for presenting term papers, final papers, scientific essays, dissertations and specialist books.

Visit us on the internet:

http://www.grin.com/

http://www.facebook.com/grincom

http://www.twitter.com/grin_com

The Triangulum

Deapon Biswas

Transport Officer, Private Concern, Chittagong, Bangladesh.

Abstract

In this paper I offered an expression widely used in this part " Formation Analysis". It is called the "Triangulum" mostly seen as a triangle. A triangulum is an expression containing components bracketed by { } where the components arranged as a triangle. It is usually denoted by $G\left\{{V \atop M}\right\}$ called triangulum of degree 'V' and width 'M'.

Key Words

Triangulum, special combination series.

Article Outline

1. Introduction
2. Triangulum
3. How to make a triangulum
4. Special combination series.
5. Conclusions

1. Introduction

I state here a definition of triangulum and 7 theorems to cover this paper. After the definition I state elaborately how to make a triangulum.

2. Triangulum

A triangulum is an expression containing components bracketed by { } where the components arranged as a triangle i.e., each row must put down one after another eliminating a component. For example

$$G\left\{ \begin{matrix} 2 \\ 5 \end{matrix} \right\} = \left\{ \begin{matrix} A & B & C & D & E \\ & F & G & H & I \\ & & J & K & L \\ & & & M & N \\ & & & & O \end{matrix} \right\}$$

where the triangulum has 5 rows and 5 columns. The component number of first row is the width of the triangulum i.e., the number of rows equal to the number of columns is the width of the triangulum. It is not essential that each row contains alike components but it is sometimes essential that each column contains alike components. The components of a triangulum may be numbers or triangulums or any other things. The last point of the upper row is called the centre of a triangulum.

3. How to make a triangulum

Let we have given a row of 6 letters A, B, C, D, E, F. Now put the 6 letters in a row; After it is done we make a second row eliminating first letter A and put the remaining 5 letters down the first row after A. After it is done we make a third row eliminating first and second letters A and B. Now put the remaining 4 letters down the second row after B. Proceeding these ways we make the 6th row containing the components F and put down the 5th row after E. The described triangulum is of degree '2' and width '6' and placed by

$$G\left\{ \begin{matrix} 2 \\ 6 \end{matrix} \right\} = \left\{ \begin{matrix} A & B & C & D & E & F \\ & B & C & D & E & F \\ & & C & D & E & F \\ & & & D & E & F \\ & & & & E & F \\ & & & & & F \end{matrix} \right\}$$

For a triangulum of degree '3' and width '6' we put the triamgulums $G\left\{ \begin{matrix} 2 \\ 6 \end{matrix} \right\}, G\left\{ \begin{matrix} 2 \\ 5 \end{matrix} \right\}, G\left\{ \begin{matrix} 2 \\ 4 \end{matrix} \right\}, G\left\{ \begin{matrix} 2 \\ 3 \end{matrix} \right\}, G\left\{ \begin{matrix} 2 \\ 2 \end{matrix} \right\}$ and $G\left\{ \begin{matrix} 2 \\ 1 \end{matrix} \right\}$ one down after another i.e.,

$$G\begin{Bmatrix}3\\6\end{Bmatrix} = \begin{Bmatrix} G\begin{Bmatrix}2\\6\end{Bmatrix} \\ G\begin{Bmatrix}2\\5\end{Bmatrix} \\ G\begin{Bmatrix}2\\4\end{Bmatrix} \\ G\begin{Bmatrix}2\\3\end{Bmatrix} \\ G\begin{Bmatrix}2\\2\end{Bmatrix} \\ G\begin{Bmatrix}2\\1\end{Bmatrix} \end{Bmatrix} \quad \text{Or,} \quad \left\{ G\begin{Bmatrix}2\\6\end{Bmatrix}, G\begin{Bmatrix}2\\5\end{Bmatrix}, G\begin{Bmatrix}2\\4\end{Bmatrix}, G\begin{Bmatrix}2\\3\end{Bmatrix}, G\begin{Bmatrix}2\\2\end{Bmatrix}, G\begin{Bmatrix}2\\1\end{Bmatrix} \right\}'$$

In the above notation the component triangulums places one right after another. The notation reads $\left\{ G\begin{Bmatrix}2\\6\end{Bmatrix}, G\begin{Bmatrix}2\\5\end{Bmatrix}, G\begin{Bmatrix}2\\4\end{Bmatrix}, G\begin{Bmatrix}2\\3\end{Bmatrix}, G\begin{Bmatrix}2\\2\end{Bmatrix}, G\begin{Bmatrix}2\\1\end{Bmatrix} \right\}$ transpose. For example A' reads A transpose. Now putting the component triangulums we get

$$G\begin{Bmatrix}3\\6\end{Bmatrix} = \left\{ \begin{matrix} A\ B\ C\ D\ E\ F \\ B\ C\ D\ E\ F \\ C\ D\ E\ F \\ D\ E\ F \\ E\ F \\ F \\ B\ C\ D\ E\ F \\ C\ D\ E\ F \\ D\ E\ F \\ E\ F \\ F \\ C\ D\ E\ F \\ D\ E\ F \\ E\ F \\ F \\ D\ E\ F \\ E\ F \\ F \\ E\ F \\ F \\ F \end{matrix} \right\} \qquad \text{————— (1)}$$

The above two triangulums $G\begin{Bmatrix}2\\6\end{Bmatrix}$ and $G\begin{Bmatrix}3\\6\end{Bmatrix}$ contain the columns are of alike components. In the same way we get $G\begin{Bmatrix}V\\M\end{Bmatrix}$; triangulum of degree 'V' and width 'M' where the columns do not contain alike components as

$$G\begin{Bmatrix} V \\ M \end{Bmatrix} = \left\{ \begin{array}{c} a_{11\ldots1111}, \quad a_{11\ldots1112}, a_{11\ldots1113}, \ldots\ldots, a_{11\ldots111M}, \\ a_{11\ldots1122}, a_{11\ldots1123}, \cdots\cdots, a_{11\ldots112M}, \\ a_{11\ldots1133}, \cdots\cdots, a_{11\ldots113M}, \\ \vdots \\ a_{11\ldots\ldots11MM}, \\ a_{11\ldots1222}, a_{11\ldots1223}, \ldots\ldots, \quad a_{11\ldots\ldots122M}, \\ a_{11\ldots1233}, \ldots\ldots, \quad a_{11\ldots\ldots123M}, \\ \vdots \\ a_{11\ldots\ldots12MM}, \\ a_{11\ldots1333}, \ldots\ldots, \quad a_{11\ldots\ldots133M}, \\ \vdots \\ a_{11\ldots\ldots13MM}, \\ \vdots \\ a_{11\ldots1MMM}, \\ a_{11\ldots2222}, a_{11\ldots2223}, \ldots\ldots, a_{11\ldots\ldots222M}, \\ a_{11\ldots2233}, \ldots\ldots, a_{11\ldots\ldots223M} \\ \vdots \\ a_{11\ldots\ldots22MM} \\ a_{11\ldots2333}, \ldots\ldots, a_{11\ldots\ldots233M} \\ \vdots \\ a_{11\ldots\ldots23MM} \\ \vdots \\ a_{11\ldots\ldots2MMM} \\ \vdots \\ a_{11\ldots\ldots1MMMM} \\ \vdots \\ a_{MM\ldots\ldots1MMMM} \end{array} \right\} \quad\text{——} \quad (2)$$

where M takes V times.

Example 1: Set the following triangulums as your choice
(i) degree 2, width 7 , (ii) degree 3, width 7 , (iii) degree 4, width 4,
(iv) degree 0, width 4 and (v) degree 1, width 4.

Solution: (i) We get for degree 2, range 7 as our choice

$$G\begin{Bmatrix}2\\7\end{Bmatrix} = \left\{\begin{array}{l} m,\ n,\ o,\ p,\ q,\ r,\ s \\ n,\ o,\ p,\ q,\ r,\ s \\ o,\ p,\ q,\ r,\ s \\ p,\ q,\ r,\ s \\ q,\ r,\ s \\ r,\ s \\ s \end{array}\right\}$$

(ii) We get for degree 3, width 7 as our choice

$$G\begin{Bmatrix}3\\7\end{Bmatrix} = \left\{\begin{array}{c} S_{111},\ S_{112},\ S_{113},\ S_{114},\ S_{115},\ S_{116},\ S_{117} \\ S_{122},\ S_{123},\ S_{124},\ S_{125},\ S_{126},\ S_{127} \\ S_{133},\ S_{134},\ S_{135},\ S_{136},\ S_{137} \\ S_{144},\ S_{145},\ S_{146},\ S_{147} \\ S_{155},\ S_{156},\ S_{157} \\ S_{166},\ S_{167} \\ S_{177} \\ S_{222},\ S_{223},\ S_{224},\ S_{225},\ S_{226},\ S_{227} \\ S_{233},\ S_{234},\ S_{235},\ S_{236},\ S_{237} \\ S_{244},\ S_{245},\ S_{246},\ S_{247} \\ S_{255},\ S_{256},\ S_{257} \\ S_{266},\ S_{267} \\ S_{277} \\ S_{333},\ S_{334},\ S_{335},\ S_{336},\ S_{337} \\ S_{344},\ S_{345},\ S_{346},\ S_{347} \\ S_{355},\ S_{356},\ S_{357} \\ S_{366},\ S_{367} \\ S_{377} \\ S_{444},\ S_{445},\ S_{446},\ S_{447} \\ S_{455},\ S_{456},\ S_{457} \\ S_{466},\ S_{467} \\ S_{477} \\ S_{555},\ S_{556},\ S_{557} \\ S_{566},\ S_{567} \\ S_{577} \\ S_{666},\ S_{667} \\ S_{677} \\ S_{777} \end{array}\right\}$$

(iii) We get for degree 4, width 4, over 1 is

$$G\left\{\begin{matrix}4\\4\end{matrix}\right\} = \left\{\begin{array}{l}1,\ 1,\ 1,\ 1\\ \quad 1,\ 1,\ 1\\ \qquad 1,\ 1\\ \qquad\quad 1\\ 1,\ 1,\ 1\\ \quad 1,\ 1\\ \qquad 1\\ 1,\ 1\\ \quad 1\\ \quad 1\\ 1,\ 1,\ 1\\ \quad 1,\ 1\\ \qquad 1\\ 1,\ 1\\ \quad 1\\ \quad 1\\ 1,\ 1\\ \quad 1\\ \quad 1\\ \quad 1\end{array}\right\}$$

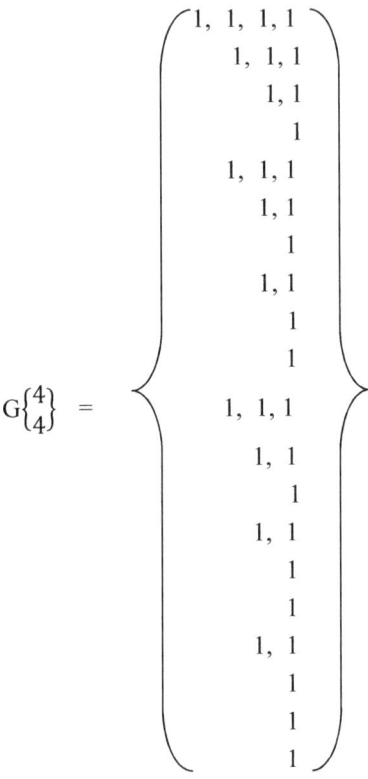

(iv) We get for degree 0, width 4, over 1 is

$$G\left\{\begin{matrix}0\\4\end{matrix}\right\} = \{A\}$$

(v) We get for degree 1, width 4, over 1 is

$$G\left\{\begin{matrix}1\\4\end{matrix}\right\} = \{E, F, G, H\}.$$

Theorem 1: The triangulum of degree V and width M can be expressed as the union of the triangulums of degree V−1 and widths 1to M i.e.,

$$G\left\{\begin{matrix}V\\M\end{matrix}\right\} = U_{i=1}^{M} G\left\{\begin{matrix}V-1\\i\end{matrix}\right\} \qquad\qquad\text{------------ (3)}$$

Proof: We know from how to make a trainingulum the triangulum $G\left\{\begin{matrix}V\\M\end{matrix}\right\}$

consists of $G\left\{\begin{matrix}V-1\\M\end{matrix}\right\}$, $G\left\{\begin{matrix}V-1\\M-1\end{matrix}\right\}$, $G\left\{\begin{matrix}V-1\\M-2\end{matrix}\right\}$, ... , $G\left\{\begin{matrix}V-1\\1\end{matrix}\right\}$. The consisting triangulums put down one after another. Thus

$$G\left\{{V \atop M}\right\} = \left\{\begin{array}{l} G\left\{{V-1 \atop M}\right\} \\ G\left\{{V-1 \atop M-1}\right\} \\ G\left\{{V-1 \atop M-2}\right\} \\ \vdots \\ G\left\{{V-1 \atop 1}\right\} \end{array}\right\}$$

$$= \left\{G\left\{{V-1 \atop M}\right\}, G\left\{{V-1 \atop M-1}\right\}, G\left\{{V-1 \atop M-2}\right\}, \ldots\ldots, G\left\{{V-1 \atop 1}\right\}\right\}'$$

As the component triangulums are mutually exclusive so we get

$$G\left\{{V \atop M}\right\} = G\left\{{V-1 \atop M}\right\} \cup G\left\{{V-1 \atop M-1}\right\} \cup G\left\{{V-1 \atop M-2}\right\} \cup \ldots \cup G\left\{{V-1 \atop 1}\right\}$$

$$= \cup_{i=1}^{M} G\left\{{V-1 \atop i}\right\}$$

Hence the proof.

Example 2: Consider the triangulum $G\left\{{3 \atop 6}\right\}$ given in (1). Now show the triangulum is the union of the component triangulums $G\left\{{2 \atop 6}\right\}, G\left\{{2 \atop 5}\right\}, G\left\{{2 \atop 4}\right\}$, $G\left\{{2 \atop 3}\right\}, G\left\{{2 \atop 2}\right\}$ and $G\left\{{2 \atop 1}\right\}$.

Solution: From the triangulum $G\left\{{3 \atop 6}\right\}$ given in (1) we get

$$G\left\{{2 \atop 6}\right\} = \left\{\begin{array}{l} A,\ B,\ C,\ D,\ E,\ F \\ B,\ C,\ D,\ E,\ F \\ C,\ D,\ E,\ F \\ D,\ E,\ F \\ E,\ F \\ F \end{array}\right\}, \quad G\left\{{2 \atop 5}\right\} = \left\{\begin{array}{l} B,\ C,\ D,\ E,\ F \\ C,\ D,\ E,\ F \\ D,\ E,\ F \\ E,\ F \\ F \end{array}\right\},$$

$$G\left\{{2 \atop 4}\right\} = \left\{\begin{array}{l} C,\ D,\ E,\ F \\ D,\ E,\ F \\ E,\ F \\ F \end{array}\right\}, \quad G\left\{{2 \atop 3}\right\} = \left\{\begin{array}{l} D,\ E,\ F \\ E,\ F \\ F \end{array}\right\}, \quad G\left\{{2 \atop 2}\right\} = \left\{\begin{array}{l} E,\ F \\ F \end{array}\right\},$$

$$G\left\{{2 \atop 1}\right\} = \{F\}$$

As the component triangulums are mutually exclusive so their union is the triangulum $G\left\{{3 \atop 6}\right\}$ i.e.,

$$G\left\{{3 \atop 6}\right\} = G\left\{{2 \atop 6}\right\} \cup G\left\{{2 \atop 5}\right\} \cup G\left\{{2 \atop 4}\right\} \cup G\left\{{2 \atop 3}\right\} \cup G\left\{{2 \atop 2}\right\} \cup G\left\{{2 \atop 1}\right\}.$$

Theorem 2: The triangulum of degree V and width M can be expressed as the triangulum of degree v width M where the components are the triangulums of degree V− v and widths 1 to M i.e.,

$$G\left\{{V \atop M}\right\} = G\left\{{V - v,\ v \atop M}\right\} \qquad\qquad\qquad (4)$$

Proof: We know from how to make a triangulum the triangulum $G\left\{{V \atop M}\right\}$ consists of $G\left\{{V - 1 \atop M}\right\}$, $G\left\{{V - 1 \atop M - 1}\right\}$, $G\left\{{V - 1 \atop M - 2}\right\}$, ... , $G\left\{{V - 1 \atop 1}\right\}$ and the consisting triangulums put down one after another i.e.,

$$G\left\{{V \atop M}\right\} = \begin{Bmatrix} G\left\{{V - 1 \atop M}\right\} \\ G\left\{{V - 1 \atop M - 1}\right\} \\ G\left\{{V - 1 \atop M - 2}\right\} \\ \vdots \\ G\left\{{V - 1 \atop 1}\right\} \end{Bmatrix}$$

$$= \left\{ G\left\{{V - 1 \atop M}\right\}, G\left\{{V - 1 \atop M - 1}\right\}, G\left\{{V - 1 \atop M - 2}\right\}, \ldots\ldots, G\left\{{V - 1 \atop 1}\right\} \right\}' = G\left\{{V - 1,\ 1 \atop M}\right\}$$

where the component triangulums are of degree V−1 and widths 1 to M. For the same reason we get

$$G\left\{{V - 1 \atop M}\right\} = \left\{ G\left\{{V - 2 \atop M}\right\}, G\left\{{V - 2 \atop M - 1}\right\}, G\left\{{V - 2 \atop M - 2}\right\}, \ldots\ldots, G\left\{{V - 2 \atop 1}\right\} \right\}'$$

$$G\left\{{V - 1 \atop M - 1}\right\} = \left\{ G\left\{{V - 2 \atop M - 1}\right\}, G\left\{{V - 2 \atop M - 2}\right\}, G\left\{{V - 2 \atop M - 3}\right\}, \ldots\ldots, G\left\{{V - 2 \atop 1}\right\} \right\}'$$

$$G\left\{{V - 1 \atop M - 2}\right\} = \left\{ G\left\{{V - 2 \atop M - 2}\right\}, G\left\{{V - 2 \atop M - 3}\right\}, G\left\{{V - 2 \atop M - 4}\right\}, \ldots\ldots, G\left\{{V - 2 \atop 1}\right\} \right\}'$$

$$\vdots$$

$$G\left\{{V - 1 \atop 1}\right\} = \left\{ G\left\{{V - 2 \atop 1}\right\} \right\}'$$

As the component triangulums are the mutually exclusive so we get

$$G\left\{{V\atop M}\right\} = \left\{\begin{array}{l} G\left\{{V-2\atop M}\right\}, G\left\{{V-2\atop M-1}\right\}, G\left\{{V-2\atop M-2}\right\}, \ldots\ldots, G\left\{{V-2\atop 1}\right\} \\ \qquad G\left\{{V-2\atop M-1}\right\}, G\left\{{V-2\atop M-2}\right\}, \ldots\ldots, G\left\{{V-2\atop 1}\right\} \\ \qquad\qquad \left\{{V-2\atop M-2}\right\}, \ \ldots\ldots, G\left\{{V-2\atop 1}\right\} \\ \qquad\qquad\qquad\qquad \vdots \\ \qquad\qquad\qquad\qquad\qquad G\left\{{V-2\atop 1}\right\} \end{array}\right.$$

$$= G\left\{{V-2,\ 2\atop M}\right\}$$

This is a triangulum of degree 2 width M where the components are the triangulums of degree V−2 and widths 1 to M. Again we get

$$G\left\{{V-2\atop M}\right\} = \left\{G\left\{{V-3\atop M}\right\}, G\left\{{V-3\atop M-1}\right\}, G\left\{{V-3\atop M-2}\right\}, \ldots\ldots, G\left\{{V-3\atop 1}\right\}\right\}'$$

$$G\left\{{V-2\atop M-1}\right\} = \left\{G\left\{{V-3\atop M-1}\right\}, G\left\{{V-3\atop M-2}\right\}, G\left\{{V-3\atop M-3}\right\}, \ldots\ldots, G\left\{{V-3\atop 1}\right\}\right\}'$$

$$G\left\{{V-2\atop M-2}\right\} = \left\{G\left\{{V-3\atop M-2}\right\}, G\left\{{V-3\atop M-3}\right\}, G\left\{{V-3\atop M-4}\right\}, \ldots\ldots, G\left\{{V-3\atop 1}\right\}\right\}'$$

$$\vdots$$

$$G\left\{{V-2\atop 1}\right\} = \left\{G\left\{{V-3\atop 1}\right\}\right\}'$$

As the component triangulums are the mutually exclusive so we get

$$
G\begin{Bmatrix} V \\ M \end{Bmatrix} = \left\{
\begin{array}{c}
G\begin{Bmatrix} V-3 \\ M \end{Bmatrix}, G\begin{Bmatrix} V-3 \\ M-1 \end{Bmatrix}, G\begin{Bmatrix} V-3 \\ M-2 \end{Bmatrix}, \ldots\ldots, G\begin{Bmatrix} V-3 \\ 1 \end{Bmatrix} \\
G\begin{Bmatrix} V-3 \\ M-1 \end{Bmatrix}, G\begin{Bmatrix} V-3 \\ M-2 \end{Bmatrix}, \ldots\ldots, G\begin{Bmatrix} V-3 \\ 1 \end{Bmatrix} \\
\begin{Bmatrix} V-3 \\ M-2 \end{Bmatrix}, \ldots\ldots, G\begin{Bmatrix} V-3 \\ 1 \end{Bmatrix} \\
\vdots \\
G\begin{Bmatrix} V-3 \\ 1 \end{Bmatrix} \\
G\begin{Bmatrix} V-3 \\ M-1 \end{Bmatrix}, G\begin{Bmatrix} V-3 \\ M-2 \end{Bmatrix}, \ldots\ldots, G\begin{Bmatrix} V-3 \\ 1 \end{Bmatrix} \\
G\begin{Bmatrix} V-3 \\ M-2 \end{Bmatrix}, \ldots\ldots, G\begin{Bmatrix} V-3 \\ 1 \end{Bmatrix} \\
\vdots \\
G\begin{Bmatrix} V-3 \\ 1 \end{Bmatrix} \\
G\begin{Bmatrix} V-3 \\ M-2 \end{Bmatrix}, \ldots\ldots, G\begin{Bmatrix} V-3 \\ 1 \end{Bmatrix} \\
\vdots \\
G\begin{Bmatrix} V-3 \\ 1 \end{Bmatrix} \\
\vdots \\
G\begin{Bmatrix} V-3 \\ 1 \end{Bmatrix}
\end{array}
\right\}
$$

$$
= G\begin{Bmatrix} V-3, & 3 \\ & M \end{Bmatrix}
$$

This is a triangulum of degree 3 width M where the components are the triangulums of degree V−3 and widths 1 to M. Proceeding these ways we get for v steps

$$
G\begin{Bmatrix} V \\ M \end{Bmatrix} = G\begin{Bmatrix} V-v, & v \\ & M \end{Bmatrix}
$$

where the right side is a triangulum of degree v and width M in which the component triangulums are of degree V−v and widths 1 to M.

Hence the proof.

Example 3: Let the triangulum $G\begin{Bmatrix} 3 \\ 6 \end{Bmatrix}$ given in (1). Now show the triangulums of the forms (i) $G\begin{Bmatrix} 2, & 1 \\ & 6 \end{Bmatrix}$ and $G\begin{Bmatrix} 1, & 2 \\ & 6 \end{Bmatrix}$.

Solution: The first row of the given triangulum $G\begin{Bmatrix} 3 \\ 6 \end{Bmatrix}$ is A, B, C, D, E, F. Now

(i) The form of $G\left\{\begin{matrix}2, & 1\\ & 6\end{matrix}\right\}$ indicates the triangulum of degree 1 and width 6 where the component triangulum are of degree 2 and widths 1 to 6. Thus we get

$$G\left\{\begin{matrix}2, & 1\\ & 6\end{matrix}\right\} = \left\{\begin{matrix} A, B, C, D, E, F & B, C, D, E, F & C, D, E, F & D, E, F & E, F & F\\ B, C, D, E, F & C, D, E, F & D, E, F & E, F & F\\ C, D, E, F & D, E, F & E, F & F\\ D, E, F & E, F & F\\ E, F & F\\ F \end{matrix}\right\}$$

(ii) The form of $G\left\{\begin{matrix}1, & 2\\ & 6\end{matrix}\right\}$ indicates the triangulum of degree 2 and width 6 where the component trinagulums are of degree 1 and widths 1 to 6. So we get

$$G\left\{\begin{matrix}1, & 2\\ & 6\end{matrix}\right\} = \left\{\begin{matrix} A, B, C, D, E, F & B, C, D, E, F & C, D, E, F & D, E, F & E, F & F\\ & B, C, D, E, F & C, D, E, F & D, E, F & E, F & F\\ & & C, D, E, F & D, E, F & E, F & F\\ & & & D, E, F & E, F & F\\ & & & & E, F & F\\ & & & & & F \end{matrix}\right\}$$

The number of components of a triangulum does not change as the degree of the triangulum may be changed.

The proof is left as an exercise.

Theorem 3: The number of components of a triangulum $G\left\{\begin{matrix}V\\M\end{matrix}\right\}$ denoted by $G\left(\begin{matrix}V\\M\end{matrix}\right)$ can be expressed as

$$G\left(\begin{matrix}V\\M\end{matrix}\right) = \sum_{i=1}^{M} G\left(\begin{matrix}V-1\\i\end{matrix}\right) \qquad\qquad (5)$$

where $G\left(\begin{matrix}V-1\\i\end{matrix}\right)$ is the number of components of the triangulum $G\left\{\begin{matrix}V-1\\i\end{matrix}\right\}$ of degree V−1 and width i.

Proof: We know the triangulum of degree V and width M can be expressed as the union of the triangulums of degreee V−1 and widths 1 to M. Thus

$$G\left\{\begin{matrix}V\\M\end{matrix}\right\} = \bigcup_{i=1}^{M} G\left\{\begin{matrix}V-1\\i\end{matrix}\right\}$$

$$= \left\{ G \left\{ \begin{matrix} V-1 \\ M \end{matrix} \right\}, G \left\{ \begin{matrix} V-1 \\ M-1 \end{matrix} \right\}, G \left\{ \begin{matrix} V-1 \\ M-2 \end{matrix} \right\}, \dots \dots, G \left\{ \begin{matrix} V-1 \\ 1 \end{matrix} \right\} \right\}'$$

Now the numbers of components of component triangulums are

for $G\left\{ \begin{matrix} V-1 \\ M \end{matrix} \right\}$ is $G\left(\begin{matrix} V-1 \\ M \end{matrix} \right)$

for $G\left\{ \begin{matrix} V-1 \\ M-1 \end{matrix} \right\}$ is $G\left(\begin{matrix} V-1 \\ M-1 \end{matrix} \right)$

for $G\left\{ \begin{matrix} V-1 \\ M-2 \end{matrix} \right\}$ is $G\left(\begin{matrix} V-1 \\ M-2 \end{matrix} \right)$

and so on

for $G\left\{ \begin{matrix} V-1 \\ 1 \end{matrix} \right\}$ is $G\left(\begin{matrix} V-1 \\ 1 \end{matrix} \right)$

Thus for $G\left\{ \begin{matrix} V \\ M \end{matrix} \right\}$ we get summing the numbers

$$G\left(\begin{matrix} V \\ M \end{matrix} \right) = G\left(\begin{matrix} V-1 \\ M \end{matrix} \right) + G\left(\begin{matrix} V-1 \\ M-1 \end{matrix} \right) + G\left(\begin{matrix} V-1 \\ M-2 \end{matrix} \right) + \dots \dots + G\left(\begin{matrix} V-1 \\ 1 \end{matrix} \right)$$

$$= \sum_{i=1}^{M} G\left(\begin{matrix} V-1 \\ i \end{matrix} \right)$$

where $G\left(\begin{matrix} V-1 \\ i \end{matrix} \right)$ is the number of components of the triangulum $G\left\{ \begin{matrix} V-1 \\ i \end{matrix} \right\}$ of degree V−1 and width i = 1, 2, 3, ……, M.

Hence the proof.

Example 4: Let the triangulum $G\left\{ \begin{matrix} 3 \\ 7 \end{matrix} \right\}$ given in example 1(ii). Now show that $G\left(\begin{matrix} 3 \\ 7 \end{matrix} \right)$ can be expressed as

$$G\left(\begin{matrix} 3 \\ 7 \end{matrix} \right) = \sum_{i=1}^{7} G\left(\begin{matrix} 2 \\ i \end{matrix} \right).$$

Solution: From the triangulum $G\left\{ \begin{matrix} 3 \\ 7 \end{matrix} \right\}$ given in example 1. We get by counting

$$G\left(\begin{matrix} 2 \\ 7 \end{matrix} \right) = 28, \ G\left(\begin{matrix} 2 \\ 6 \end{matrix} \right) = 21, \ G\left(\begin{matrix} 2 \\ 5 \end{matrix} \right) = 15, \ G\left(\begin{matrix} 2 \\ 4 \end{matrix} \right) = 10, \ G\left(\begin{matrix} 2 \\ 3 \end{matrix} \right) = 6, \ G\left(\begin{matrix} 2 \\ 2 \end{matrix} \right) = 3$$

and $G\left(\begin{matrix} 2 \\ 1 \end{matrix} \right) = 1$

and for $G\left(\begin{matrix} 3 \\ 7 \end{matrix} \right)$ by counting

$$G\left(\begin{matrix} 3 \\ 7 \end{matrix} \right) = G\left(\begin{matrix} 2 \\ 7 \end{matrix} \right) + G\left(\begin{matrix} 2 \\ 6 \end{matrix} \right) + G\left(\begin{matrix} 2 \\ 5 \end{matrix} \right) + G\left(\begin{matrix} 2 \\ 4 \end{matrix} \right) + G\left(\begin{matrix} 2 \\ 3 \end{matrix} \right) + G\left(\begin{matrix} 2 \\ 2 \end{matrix} \right) + G\left(\begin{matrix} 2 \\ 1 \end{matrix} \right)$$

$$= 28 + 21 + 15 + 10 + 6 + 3 + 1 = 84.$$

Theorem 4: The number of components of a triangulum $G\begin{Bmatrix}V\\M\end{Bmatrix}$ denoted by $G\begin{pmatrix}V\\M\end{pmatrix}$ can be expressed as the sum of numbers of components of the triangulum $G\begin{Bmatrix}V-v,\ v\\M\end{Bmatrix}$; denoted by $S_vG\begin{pmatrix}V-v\\i\end{pmatrix}_{i=1\ to\ M}$ (read sum to degree v of $G\begin{pmatrix}V-v\\i\end{pmatrix}_{i=1\ to\ M}$) i.e.,

$$G\begin{pmatrix}V\\M\end{pmatrix} = S_vG\begin{pmatrix}V-v\\i\end{pmatrix}_{i=1\ to\ M} \makebox[3cm]{\hrulefill} (6)$$

Proof: We know the number of components of a triangulum does not change as the degree of the triangulum may be changed i .e.,

$$G\begin{pmatrix}V\\M\end{pmatrix} = G\begin{pmatrix}V-v,\ v\\M\end{pmatrix}$$

Now the number of components of the triangulum $G\begin{Bmatrix}V-v,\ v\\M\end{Bmatrix}$ denoted by $G\begin{pmatrix}V-v,\ v\\M\end{pmatrix}$ is the same as the sum of numbers of components of every component triangulums. Hence we get

$$G\begin{pmatrix}V\\M\end{pmatrix} = S_vG\begin{pmatrix}V-v\\i\end{pmatrix}_{i=1\ to\ M}$$

Example 5: Find the number of components of the triangulum $G\begin{Bmatrix}3\\6\end{Bmatrix}$ given in form $G\begin{Bmatrix}2,\ 1\\6\end{Bmatrix}$ and $G\begin{Bmatrix}1,\ 2\\6\end{Bmatrix}$ shown in example 3.

Solution: (i) The form of $G\begin{Bmatrix}2,\ 1\\6\end{Bmatrix}$ indicates the triangulum of degree 1 and width 6 where the component triangulums are of degree 2 and widths 1 to 6. So the number of components contained in $G\begin{Bmatrix}2,\ 1\\6\end{Bmatrix}$ is the same as the sum of numbers of components of component triangulums contained in $G\begin{Bmatrix}2,\ 1\\6\end{Bmatrix}$. Hence

$$G\begin{pmatrix}3\\6\end{pmatrix} = G\begin{pmatrix}2,\ 1\\6\end{pmatrix}$$

$$= S_1G\begin{pmatrix}2\\i\end{pmatrix}_{i=1\ to\ 6}$$

$$= S_1\left[G\begin{pmatrix}2\\6\end{pmatrix}, G\begin{pmatrix}2\\5\end{pmatrix}, G\begin{pmatrix}2\\4\end{pmatrix}, G\begin{pmatrix}2\\3\end{pmatrix}, G\begin{pmatrix}2\\2\end{pmatrix}, G\begin{pmatrix}2\\1\end{pmatrix}\right]$$

$$= \left[G\begin{pmatrix}2\\6\end{pmatrix} + G\begin{pmatrix}2\\5\end{pmatrix} + G\begin{pmatrix}2\\4\end{pmatrix} + G\begin{pmatrix}2\\3\end{pmatrix} + G\begin{pmatrix}2\\2\end{pmatrix} + G\begin{pmatrix}2\\1\end{pmatrix}\right]$$

13

$$= [21 + 15 + 10 + 6 + 3 + 1]$$
$$= 56.$$

(ii) The form of $G\{^{1,\ 2}_{\ \ 6}\}$ indicates the triangulum of degree 2 and width 6 where the component triangulums are of degree 1 and widths 1 to 6. Thus the number of components contained in $G\{^{1,\ 2}_{\ \ 6}\}$ is the same as the sum of numbers of components of component triangulums contained in $G\{^{1,\ 2}_{\ \ 6}\}$.

Hence

$$G\binom{3}{6} = G\binom{1,\ 2}{6}$$

$$= S_2 G\binom{1}{i}_{i=1 \text{ to } 6}$$

$$= S_2\left[G\binom{1}{6}, G\binom{1}{5}, G\binom{1}{4}, G\binom{1}{3}, G\binom{1}{2}, G\binom{1}{1}\right]$$

$$=
\begin{bmatrix}
G\binom{1}{6} + G\binom{1}{5} + G\binom{1}{4} + G\binom{1}{3} + G\binom{1}{2} + G\binom{1}{1} \\
+ G\binom{1}{5} + G\binom{1}{4} + G\binom{1}{3} + G\binom{1}{2} + G\binom{1}{1} \\
+ G\binom{1}{4} + G\binom{1}{3} + G\binom{1}{2} + G\binom{1}{1} \\
+ G\binom{1}{3} + G\binom{1}{2} + G\binom{1}{1} \\
+ G\binom{1}{2} + G\binom{1}{1} \\
+ G\binom{1}{1}
\end{bmatrix}$$

$$=
\begin{bmatrix}
6 + 5 + 4 + 3 + 2 + 1 \\
+5 + 4 + 3 + 2 + 1 \\
+4 + 3 + 2 + 1 \\
+3 + 2 + 1 \\
+2 + 1 \\
+1
\end{bmatrix} = 56.$$

Theorem 5: The number of components of a tirangulum $G\{^{V}_{M}\}$ denoted by $G\binom{V}{M}$ can be expressed as

$$G\binom{V}{M} = \frac{1}{2}\left[\text{SNG}\binom{V-2}{M} + \text{SNG}\binom{V-2}{M^2}\right] \quad\quad\quad (7)$$

14

where $\mathrm{SNG}\left(\begin{smallmatrix} V-2 \\ M \end{smallmatrix}\right)$ denotes sum of numbers of triangulum of natural numbers and $\mathrm{SNG}\left(\begin{smallmatrix} V-2 \\ M^2 \end{smallmatrix}\right)$ denotes sum of numbers of triangulum of natural number squares.

Proof: Consider the triangulum $G\left\{\begin{smallmatrix} 2 \\ M \end{smallmatrix}\right\}$. Then by definition we get the number of components is

$$G\left(\begin{smallmatrix} 2 \\ M \end{smallmatrix}\right) = \frac{M^2-M}{2} + M = \frac{1}{2}[M + M^2] = \frac{1}{2}\left[\mathrm{SNG}\left(\begin{smallmatrix} 0 \\ M \end{smallmatrix}\right) + \mathrm{SNG}\left(\begin{smallmatrix} 0 \\ M^2 \end{smallmatrix}\right)\right]$$

Again consider the triangulum $G\left\{\begin{smallmatrix} 3 \\ M \end{smallmatrix}\right\}$. Then by definition we get the number of components is

$$G\left(\begin{smallmatrix} 3 \\ M \end{smallmatrix}\right) = \frac{M^2-M}{2} + M + \frac{(M-1)^2-(M-1)}{2} + (M-1) + \frac{(M-2)^2-(M-2)}{2} + (M-2)$$
$$+ \ldots + \frac{1^2-1}{2} + 1$$
$$= \frac{1}{2}[\{M + (M-1) + (M-2) + \ldots + 1\} + \{M^2 + (M-1)^2$$
$$+ (M-2)^2 + \ldots + 1^2\}]$$
$$= \frac{1}{2}\left[\mathrm{SNG}\left(\begin{smallmatrix} 1 \\ M \end{smallmatrix}\right) + \mathrm{SNG}\left(\begin{smallmatrix} 1 \\ M^2 \end{smallmatrix}\right)\right]$$

Now consider the triangulum $G\left\{\begin{smallmatrix} 4 \\ M \end{smallmatrix}\right\}$. Then by definition we get the number of components is

$$= G\left(\begin{smallmatrix} 4 \\ M \end{smallmatrix}\right) = \frac{M^2-M}{2} + M + \frac{(M-1)^2-(M-1)}{2} + (M-1) + \frac{(M-2)^2-(M-2)}{2} + (M-2)$$
$$+ \ldots + \frac{1^2-1}{2} + 1 + \frac{(M-1)^2-(M-1)}{2} + (M-1) + \frac{(M-2)^2-(M-2)}{2} + (M-2) + \ldots +$$
$$\frac{1^2-1}{2} + 1 + \frac{(M-2)^2-(M-2)}{2} + (M-2) + \ldots + \frac{1^2-1}{2} + 1 + \ldots + \frac{1^2-1}{2} + 1$$
$$= \frac{1}{2}[\{M + (M-1) + (M-2) + \ldots + 1\} + \{M^2 + (M-1)^2 + (M-2)^2 +$$
$$\ldots + 1^2\} + \{(M-1) + (M-2) + \ldots + 1\} + \{(M-1)^2 + (M-2)^2 + \ldots + 1^2\}$$
$$+ \{((M-2) + \ldots + 1\} + \{(M-2)^2 + \ldots + 1^2\} + \ldots + \{1\} + \{1^2\}]$$
$$= \frac{1}{2}\left[\mathrm{SNG}\left(\begin{smallmatrix} 2 \\ M \end{smallmatrix}\right) + \mathrm{SNG}\left(\begin{smallmatrix} 2 \\ M^2 \end{smallmatrix}\right)\right]$$

Proceeding these ways we get for the triangulum $G\left(\begin{smallmatrix} V \\ M \end{smallmatrix}\right)$ the number of components is

$$G\left(\begin{smallmatrix} V \\ M \end{smallmatrix}\right) = \frac{1}{2}\left[\mathrm{SNG}\left(\begin{smallmatrix} V-2 \\ M \end{smallmatrix}\right) + \mathrm{SNG}\left(\begin{smallmatrix} V-2 \\ M^2 \end{smallmatrix}\right)\right]$$

Example 6: Using theorem 5 find the number of components of the following triangulums (i) $G\{^4_6\}$ and (ii) $G\{^5_6\}$.

Solution: (i) The number of components of the triangulum $G\{^4_6\}$ is given by

$$G\binom{4}{6} = \frac{1}{2}\left[\text{SNG}\binom{2}{6} + \text{SNG}\binom{2}{6^2}\right] = \frac{1}{2}[56 + 196] = 126.$$

(ii) The number of components of the triangulum $G\{^5_6\}$ is given by

$$G\binom{5}{6} = \frac{1}{2}\left[\text{SNG}\binom{3}{6} + \text{SNG}\binom{3}{6^2}\right] = \frac{1}{2}[126 + 378] = 252.$$

Theorem 6: The number of components of a triangulum $G\{^V_M\}$ is the same as the number of formations of M sided V dice experiment i.e.,

$$G\binom{V}{M} = F\binom{M}{V} \qquad\qquad\qquad\text{————————————— (8)}$$

Proof: Let the triangulum

$$G\{^0_M\} = \{M\}$$

Then the number of components of the triangulum is

$$G\binom{0}{M} = 1 = C\binom{M+0-1}{0} = F\binom{M}{0}$$

Again let the triangulum

$$G\{^1_M\} = \{M, (M-1), (M-2), \ldots\ldots, 1\}$$

Then the number of components of the triangulum is

$$G\binom{1}{M} = M-1+1 = M = C\binom{M+1-1}{1} = F\binom{M}{1}$$

Now let the triangulum

$$G\{^2_M\} = \begin{cases} M, (M-1), (M-2), \ldots\ldots, 1 \\ (M-1), (M-2), \ldots\ldots, 1 \\ (M-2), \ldots\ldots, 1 \\ \vdots \\ 1 \end{cases}$$

Then the number of components of the triangulum is

$$G\binom{2}{M} = M + (M{-}1) + (M{-}2) + \ldots \ldots + 1 = \frac{M(M+1)}{2!} = C\binom{M+2-1}{2}$$

$$= F\binom{M}{2}$$

Similarity for the riangulum $G\left\{\dfrac{3}{M}\right\}$ we get the number of components

$$G\binom{3}{M} = M + (M{-}1) + (M{-}2) + \ldots\ldots + 1$$
$$+ (M{-}1) + (M{-}2) + \ldots\ldots + 1$$
$$+ (M{-}2) + \ldots\ldots + 1$$
$$\vdots$$
$$+ 1$$

$$= \tfrac{1}{2}M(M+1) + \tfrac{1}{2}(M{-}1)M + \tfrac{1}{2}(M{-}2)(M{-}1) + \ldots\ldots + \tfrac{1}{2}1.2$$

Let the general term of the series is

$$\tfrac{1}{2}M(M+1) = \tfrac{1}{2}(M^2 + M)$$

Now the sum of the series is

$$S = \frac{1}{2}\left\{\frac{M(M+1)(2M+1)}{6} + \frac{M(M+1)}{2}\right\} = \frac{M(M+1)(M+2)}{3!}$$

Thus we get

$$G\binom{3}{M} = \frac{M(M+1)(M+2)}{3!} = C\binom{M+3-1}{3} = F\binom{M}{3}$$

Proceeding these ways we get for the triangulum $G\left\{\dfrac{V}{M}\right\}$ the number of components is

$$G\binom{V}{M} = \frac{M(M+1)(M+2)\ldots\ldots(M+V-1)}{V!} = C\binom{M+V-1}{V} = F\binom{M}{V}$$

Hence the proof.

Example 7: Using the theorem 6 find the number of components of the following triangulums (i) $G\left\{\dfrac{4}{7}\right\}$ and (ii) $G\left\{\dfrac{6}{8}\right\}$.

Solution: (i) From the theorem 5.6 we get

$$G\binom{4}{7} = F\binom{7}{4} = C\binom{7+4-1}{4} = C\binom{10}{4} = 210.$$

(ii) From the theorem 5.6 we get

$$G\binom{6}{8} = F\binom{8}{6} = C\binom{8+6-1}{6} = C\binom{13}{6} = 1716.$$

4. Special combination series

Theorem 7: The number of combinations of M objects taken V at a time can be expressed as

$$C\binom{M}{V} = \sum_{i=V-1}^{M-1} C\binom{i}{V-1} \qquad\qquad\qquad (9)$$

Proof: We know from theorem 3 the number of components of a triangulum $G\{^V_M\}$ can be expressed as

$$G\binom{V}{M'} = \sum_{i=1}^{M'} G\binom{V-1}{i}$$
$$= G\binom{V-1}{1} + G\binom{V-1}{2} + G\binom{V-1}{3} + \dots + G\binom{V-1}{M'}$$

Now from theorem 5.6 we get

$$F\binom{M'}{V} = F\binom{1}{V-1} + F\binom{2}{V-1} + F\binom{3}{V-1} + \dots + F\binom{M'}{V-1}$$
$$\Rightarrow C\binom{M'+V+1}{V} = C\binom{V-1}{V-1} + C\binom{V}{V-1} + C\binom{V+1}{V-1} + \dots$$
$$+ C\binom{M'+V-1}{V-1}$$

Taking M'+ V−1 = M we get

$$C\binom{M}{V} = C\binom{V-1}{V-1} + C\binom{V}{V-1} + C\binom{V+1}{V-1} + \dots + C\binom{M-1}{V-1}$$
$$= \sum_{i=V-1}^{M-1} C\binom{i}{V-1}$$

Hence the proof.

Example 8: Using the theorem expend the following numbers of combinations into special combination series (i) $C\binom{8}{3}$ and (ii) $C\binom{9}{5}$.

Solution: (i) We get from special combination series

$$C\binom{8}{3} = \sum_{i=2}^{7} C\binom{i}{2} = C\binom{2}{2} + C\binom{3}{2} + C\binom{4}{2} + C\binom{5}{2} + C\binom{6}{2} + C\binom{7}{2}$$

(ii) We get from special combination series

$$C\binom{9}{5} = \sum_{i=4}^{8} C\binom{i}{4} = C\binom{4}{4} + C\binom{5}{4} + C\binom{6}{4} + C\binom{7}{4} + C\binom{8}{4}$$

5. Conclusions

Triangulum can be used in the analysis of combinations, permutations, formations and homogenations. This topic may be included in school and college algebra.

References

1. Deapon Biswas, Paper 6, Summation methods Bystematics My Classic, 2010 Self published, Chittagong, 2016 Monon Prokashon, Chittagong, Bystematics Vol. I, My Classic, 2018 Scholar's Press EU, ISBN: 987- 620-2-30664-5.

2. Deapon Biswas, Paper 13, On the combinations, Bystematics My Classic, 2010 Self published, Chittagong, 2016 Monon Prokashon, Chittagong, Bystematics Vol. II, My Classic, 2018 Scholar's Press EU, ISBN: 987- 620-2-30960-8.

3. Deapon Biswas, Paper 15, B series, Bystematics My Classic, 2010 Self published, Chittagong, 2016 Monon Prokashon, Chittagong, Bystematics Vol. II, My Classic, 2018 Scholar's Press EU, ISBN: 987- 620-2-30960-8.

4. Deapon Biswas, Paper 18, Fomrations, Bystematics My Classic, 2010 Self published, Chittagong, 2016 Monon Prokashon, Chittagong, Bystematics Vol. II, My Classic, 2018 Scholar's Press EU, ISBN: 987- 620-2-30960-8.

5. Deapon Biswas, Paper 19, Triangulum, Bystematics My Classic, 2010 Self published, Chittagong, 2016 Monon Prokashon, Chittagong, Bystematics Vol. II, My Classic, 2018 Scholar's Press EU, ISBN: 987- 620-2-30960-8.

6. F. Mosteller, R. E. K Rourke & G. B. Thomas Jr. Probability with Statistical Applications.

7. S. C Gupta & V. K. Kapoor. Fundamental of Mathemetical Statistics.

YOUR KNOWLEDGE HAS VALUE

- We will publish your bachelor's and master's thesis, essays and papers

- Your own eBook and book - sold worldwide in all relevant shops

- Earn money with each sale

Upload your text at www.GRIN.com
and publish for free